Iheb ESSOUSSI

Guidelines for good cultivation practice in forest nurseries

Iheb ESSOUSSI

Guidelines for good cultivation practice in forest nurseries

ScienciaScripts

This book is a translation from the original published under ISBN 978-620-2-54953-0.

Publisher:
Sciencia Scripts
is a trademark of
International Book Market Service Ltd., member of OmniScriptum Publishing Group
17 Meldrum Street, Beau Bassin 71504, Mauritius
Printed at: see last page
ISBN: 978-620-3-25730-4

Table of Contents

List of figures

List of tables

Introduction :

The young seedlings undergo a monitoring and control program from sowing to transplanting in the reforestation site to finally have plants that meet the standards and also the objectives set (plants suitable for reforestation; i.e. capable of withstanding the difficult conditions they will encounter later on in the field). Indeed, the seedlings from the nurseries have the capacity to survive and resist better than seeds sown directly in place or by natural regeneration. This is why nursery stock is used as planting material for production, protection or recreational plantings. This work is very meticulous, delicate and expensive, that's why it requires a quality workforce so that it can control and follow the whole process at all times.

1- Choice of species :

Before each reforestation project, the forester must first determine the species to be reforested, so that the species chosen will be successful in terms of climate and ecology.

It is obvious that the forest species endemic to such a region are more suitable for a reforestation project since it is linked to the ecological requirements of the species (climate, temperature, soil...). However, this particularity increases the **risk of spreading diseases** or **parasites over large areas,** which consequently reduces the life expectancy of the trees planted, which makes it essential to establish accompanying work to overcome any reforestation failure.

Moreover, the choice of species also poses problems that are not only ecological but also economic, technological and physiological.

- ❖ **Economic and technological criteria:** What is essential is to make species with a good technological and germinative faculty (capacity) that will be able to meet current or future consumption needs.

- ❖ **Ecological and physiological criteria:** Species are selected according to their capacity to regenerate, their taking after planting, their physiological aspect, their growth and their ease of silvicultural management. It is preferable to choose species by taking into account their coniferous or deciduous characteristics.

2- Harvest and conservation of seeds :

2.1- Seeds harvesting :

2.1.1- Origin of the seeds or the seed carrier :

Another section that judges the affinities of reforestation (successful, failed) is the **seed** and its **origin**. Indeed the **seed carriers** must be **healthy, vigorous, adult** (free of parasitic attacks: insects or fungi), having the **best possible configuration,** and especially **close to the place of future use** (adaptation to local conditions).

2.1.2- Harvesting techniques :

It seems necessary, that the identification of the seed carriers is indispensable in advance. The collection of seeds from trees that meet the standards. For this purpose, mechanical machines **(shakers, shakers, scorers, etc.)** similar to those used in agriculture to collect fruits.

Tests have shown that this mechanization was of fourfold interest:

> ➢ Speed of the operation.
> ➢ Possibility of harvesting on large trees ;
> ➢ Obtaining healthier seeds ;
> ➢ Replacement of a workforce in case of scarcity.

2.1.3- Seed quality :

The seeds that are harvested are of good quality:

> ✓ Adaptation to climatic conditions: setting (putting the fruit) in cold conditions, resistance to virescence (peeling) and cracking ;
> ✓ It must have good productivity and genetic resistance;
> ✓ It must be free of parasites and pathogens;
> ✓ It must have a high percentage of germination;
> ✓ It must be accompanied by a note indicating the scientific name of the species, place and date of harvest, lot number.

2.1.4- Harvest period :

Before thinking (considering) about harvesting the seeds, they must be harvested at the **right time when they** are **fully mature and able to germinate** and **produce young seedlings.**

2.2- Conservation of seeds :

2.2.1- Basic Principle :

Seed saving is a tool used to **store seeds for a long period of time "if the standard storage conditions" to** have a **controlled seed bank is** used in the implementation of forest reforestation programs.

2.2.2- Why conserve seeds?

The conservation of seeds is an intuitive gesture on the part of the forester, given the interests behind this operation. First of all, the forester has set up a **seed** bank whose **origins** and **carriers are known,** thus avoiding the importation of seeds from elsewhere whose **origin** and **storage conditions are ambiguous.** All these reasons turn on :

- The success of the reforestation program ;

- To have a seine stand, free of diseases ;

- Meets the objectives of its implementation;

- Obtaining a stand has a valuable economic value (wood product, not wood product).

2.2.3- Storage conditions :

The conservation of seeds, requires that the seeds are first **dried** carefully in the **shade** and in the **open air before storage**; otherwise **the seeds** are **likely to be a reservoir of germs and moisture, and later** become **useless.**

The storage place must be **dry,** i.e., **less ambient humidity**, and **no source of moisture, and cool** (maximum temperature does not **exceed 10°C), the cooler the temperature, the longer the seeds stay alive,** no heat source nearby (e.g., boiler).

2.2.3.1- Store the seeds in suitable containers :

The conservation of forest seeds does not pose any particular difficulties even **if** it lasts several years, **if storage conditions are favourable.** It **seems necessary** that the seeds should be stored in **airtight containers placed away from the light,** to be ready for sowing next year. The airtight containers are containers that isolate light and humidity.

It is possible to use **simple paper envelopes,** which take up little space and are easy to store, but **their disadvantage is that they are not hermetically sealed.** In this case, it is essential to place it in an airtight metal box. It is also **preferable to use** paper envelopes that are **not bleached with chlorine (a substance that is harmful to the quality of the seed).**

2.2.3.2- Labeling the seed boxes:

To better know the origin of the seeds intended for reforestation, the **traceability of** seeds seemed necessary. The traceability consists in a label placed on each box containing the seeds, which is called **labelling.** The label contains all the necessary information that the silviculturist in general or the reforester in particular must know. Among this information is the following:

- Name of the forest ;

- The origin (to know the environmental conditions: ecological, climatic, edaphic...);

- The lot or parcel number ;

- The name of the species ;

- Variety, if applicable ;

- The date of harvest: (to know the season of harvest, in other words is what the seeds are well matured).

3.   Cultivation of plants :

3.1-   Seed pre-treatment :

The seeds of many tree species germinate without difficulty when placed in ideal humidity and temperature conditions; others go through a dormant phase. When this dormancy is strong, artificial regeneration requires pre-treatment, which is the only way to ensure a high germination rate in a very short and regular time. Pre-treatment methods vary according to the type of seed dormancy. The main types are as follows: Integumentary, embryonic and combined (both integumentary and embryonic) dormancy.

3.1.1- Exogenous dormancy (integumentary) :

The seeds of some species have hard teguments, which totally prevent the imbibition of water and sometimes even gas exchanges. However, without imbibition and gaseous exchanges. In order to resume normal growth, several methods are used:

3.1.1.1- Physical method :

One of the simplest and most direct physical methods is to **cut, drill or file the** seed **coat of** each seed before sowing, in order to make a small hole.

Scarification (scratches) can be reinforced by soaking in cold water before sowing. If large quantities of seed are to be treated, mechanical scarification is preferable to the manual method. Seeds can be mixed in a concrete mixer with sharp-edged gravel or sand, or in a special drum lined with abrasive material (sandpaper, cement, crushed glass, etc.) or with abrasive discs. If sand or gravel is used, it should be sieved, so that it can be easily separated from the seed with a suitable mesh sieve.

3.1.1.2- Soaking in water :

Consists of soaking the seeds in water. This wet treatment combines the effects of softening of hard seed coat and leaching of chemical inhibitors.

Some seeds with low resistance to germination react favourably to soaking for 24 hours in water at room temperature. This can be explained by a faster soaking than in a wet seedbed. This treatment is recommended after manual, mechanical or chemical scarification of certain species.

3.1.1.3- Acid treatment :

The chemical most frequently used to lift the integumentary dormancy is **concentrated sulfuric acid (H_2SO_4).** This treatment is, for some species, more effective than hot water treatment. Seeds stored for a long time in the store usually require a longer immersion in acid than fresh seeds, which would not be resistant to a treatment of this duration. The handling of sulphuric acid requires the utmost care and cannot be left to inexperienced workers. The materials and equipment required are as follows:

- ✓ Commercial grade sulfuric acid (density 1.84, 95% purity)
- ✓ Acid-resistant containers (preferably made of thick plastic)
- ✓ Wire containers and sieves for handling, draining and washing of seeds ;
- ✓ Plentiful running water ;
- ✓ A safe place to drain off the diluted acid resulting from rinsing the seeds.
- ✓ Facilities for drying seeds after rinsing.

The different stages of acid treatment are as follows:

1. Allow the seeds to reach the air temperature. If they have been stored cold, do not open the container until complete temperature equilibrium has been reached. Condensation will form on cold seeds exposed to hot, humid air and may react with the acid, raising the temperature to the critical value.
2. Carefully mix the seeds to be treated in a batch.
3. Immerse the seeds in acid for the required period of time, making sure that all are well covered. The treatment temperature should be between 18 °C and 27 °C, preferably towards the upper limit of this range. A low temperature requires a longer soaking time. Careful stirring shortens the treatment.
4. Remove the seeds from the acid and wash them promptly and thoroughly under cold running water for 5 to 10 minutes to remove all traces of acid. Rinse thoroughly at first and stir the seeds carefully throughout the rinsing process.
5. Spread the seeds in a thin layer to dry, unless you prefer to sow them while still wet.

3.1.1.4- Biological methods :

In nature, animals and microorganisms play an important role in restoring integumentary permeability. It is difficult to use them for controlled seed pre-treatment, but sometimes good results have been achieved.

The seeds of **Acacia Senegal and Silica Ceratonia** that have passed through the digestive tract of goats germinate easily when the conditions are right, as a result of the action of the powerful digestive juices. A convenient pre-treatment of these species is therefore to feed pods to goats kept in pens and to recover the seeds from their droppings.

3.1.1.5- Dry heat and fire :

Solar radiation alone is not a treatment that can promote germination, but it is an important part of the alternate wetting and drying treatment described in the section "Soaking in Water".

3.1.2- Endogenous dormancy (embryonic) :

Embryonic dormancy is an essential characteristic of certain genera in temperate regions. Endogenous dormancy occurs in both **recalcitrant** and **orthodox seeds.**

- **Orthodox seeds :** Desiccation tolerant (highly dehydrated; easily survive in a dehydrated state).
- **Recalcitrant seeds :** Intolerant to desiccation (rich in water; die if they dehydrate.

Endogenous dormancy includes the case of embryos whose physiological development is incomplete at the time of release or harvest and which do not germinate until the embryos have reached maturity **(morphological dormancy).**

Most species with incompletely developed embryos also show **physiological dormancy, which means that** wet heat treatment must be supplemented with wet cold treatment.

3.1.2.1- Morphological dormancy survey :

The period of pre-treatment with **moist heat** is necessary before the embryos are sufficiently developed to be able to germinate.
1. Soak the seeds for 48 hours in several times their volume of cold water (about 3 to 5 °C).
2. Drain the seeds, then mix them with two to four times their volume of a moistened hydrophilic substance, such as sand, sand and peat mixture or vermiculite.
3. Store at fairly high temperature. A constant temperature of 20 to 25°C or an alternating temperature of 20 to 30°C is suitable for most species.
4. Open the containers weekly, stir the seeds and, if there are signs of surface dryness, moisten again by spraying with water.

3.1.2.2- Lifting of physiological dormancy :
3.1.2.2.1- Cold stratification :

Strictly speaking, stratification refers to the method of arranging seeds in alternating layers of a water-retaining medium, such as sand, peat or vermiculite, and keeping them at a low temperature of 1 to 5°C for a period of usually 20 to 60 days, but which varies considerably from species to species. The combination of high humidity and low temperature apparently triggers biochemical changes that transform complex nutrients into simpler substances that can be directly assimilated by the embryo when it starts growing again at the time of germination.

To be successful, stratification and moisture pre-cooling must meet three main conditions: a renewable source of moisture for the seed, low temperature and adequate aeration.

3.1.2.2.2- Chemical treatment :

From chemicals to tests to evaluate their ability to interrupt **endogenous dormancy**. These include **gibberellic acid, citric acid, hydrogen peroxide** and a number of other compounds. Some have shown fairly good results.

Gibberellic acid is effective for the dormancy raising in eucalyptus trees (*of E. delegatensis, E. fastigata and E. regnans*).

3.1.2.2.3- Other treatments :

X-rays, gamma rays, light radiation in the red part of the spectrum and high frequency sound waves have all been used experimentally to lift dormancy and stimulate germination.

3.1.3- Combined sleep :

Some species combine several forms of dormancy at the same time. A pre-treatment to remove one kind of dormancy is largely ineffective if it is not followed by a second pre-treatment to remove the other kind.

Quite often the physical dormancy of the integument is combined with the physiological dormancy of the embryo. In this case, the integumentary dormancy must first be treated, for example by scarification, and then pre-cooling with moist cold in order to lift the embryonic dormancy.

Each individual treatment results in less than 10 percent germination, whereas successive application of treatments against seed coat and endogenous dormancy results in rapid germination of 45 percent (mechanical scarification) or 65 percent (acid scarification) of the seed, **e.g. *Cercis canadensis*.**

3.2- Germination :

After the dormancy survey. The seeds are ready to germinate and able to give young seedlings. Germination is the stage by which a living seed slows down "wakes up" and gives birth to a seedling. This stage involves complex physiological mechanisms that are fairly well identified today. In 1957, Evenari proposes the following definition: germination is a process whose limits are the beginning of the hydration of the seed and the very beginning of the growth of the radicle.

This definition, adopted by physiologists, is validated by measurements of imbibition and respiratory activity carried out on germinating seeds. It is thus demonstrated that germination comprises three successive phases (figure 1): the imbibition phase, the germination phase stricto sensu and the growth phase. We find these same three stages for the respiratory activity.

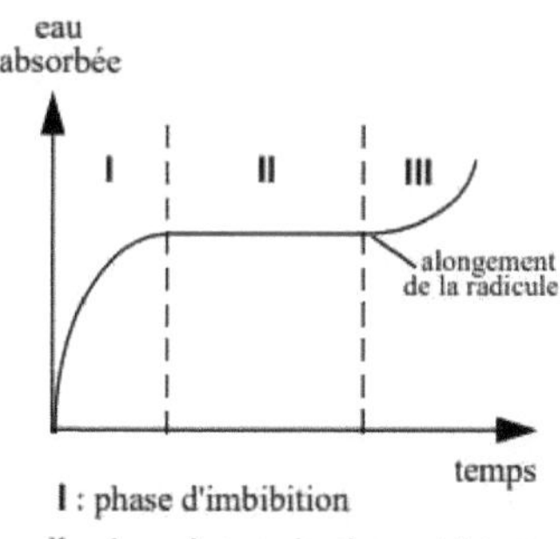

Figure 1: Theoretical seed imbibition curve (after Como, 1982).

3.3- Nursery treatment :

To sterilize the nursery against pathogens, a 40 per cent solution of "formaldehyde" applied at a rate of 80 CC per 5 liters of water is used 7 to 10 days before planting. For soil disinfection, "Methyl Bromide" is no longer used and is replaced by (DD90 "Dichloropropene Dipropene 90").

3.4- Stuffing of containers :
3.4.1- Characteristics of the containers :

The containers that are used for the cultivation of plants in nurseries are **multi-pot** containers. In terms of colors, preference has been given to opaque black plastic bags whose color seems to limit the proliferation of algae on the walls of the bag, but it is not obvious that the algae influence the good arrival of the plants. What further penalizes the transparent pot is that it is penetrated by ultraviolet radiation, which leads to molecular decomposition.

The pots, are composed of **15** toasted **cells,** used to drain excess water, **rounded** shape **and molded** between them by the upper end (a single block). The lower surface of the cells is smaller than the upper part, which gives a conical shape to these containers.

The size of the pots are designed in a way that makes them easy to use. The length is 34.5 cm, width is 22 cm, and height is 12.5 cm.

3.4.2- Stuffing of containers :

Before starting to fill the containers, it is first necessary to know the germination percentage of the seed. The germination rate must be known as soon as they are received. A high germination rate facilitates the work of the nurseryman who will consequently reduce the costs of the specific operations for the transplanting of the seedlings. This germination test must have been carried out less than a year in advance.

Moreover, cleaning the containers is a very substantial task to get the expected results. To avoid any form of pathological or entomological contamination, the nurseryman is called to clean the containers before sowing with water to remove weeds and roots, putting chloric water (bleach) to ensure disinfection against each disease. In this way the quality of the pots also has to be taken into consideration. It is necessary to clean the surroundings of the lot which are considered a host for certain wood-boring or deflating insects. In the last step the containers are filled once through the substrate.

Figure 2: Containers used in nurseries

Generally speaking, the filling of the containers is done manually.

- ✓ First fill halfway and then shake to compact;
- ✓ Fill completely and use fingers to lightly compact the cavities, especially those located at the corner of the containers;
- ✓ Shake one last time and remove excess substrate by hand;
- ✓ Monitor for proper and uniform compaction ;
- ✓ The stacking of containers should be staggered to avoid excessive compaction;
- ✓ Drop test the container at a height of about 40 cm to assess compaction (if compaction is adequate, then there will be a slight sag of 1 cm).

3.4.3- The substrate :

The substrate is the first medium where the semi or the young seedling begins to live, for this reason it must meet the needs of the plant. The soil or substrate of sowing must be light and free of diseases. It is better to make the nursery in boxes with a substrate containing 50% sand and 50% compost or a good quality compost. A good quality substrate should have the following criteria :

- ✓ Good porosity to ensure root development.
- ✓ Low soluble salt content
- ✓ Good richness in organic matter.
- ✓ A cation exchange capacity of 10 to 30 meq/100g.
- ✓ Ph of the water must be neutral, between 5.5 and 6.5

3.5-   Seeding of containers :

The success of a seedling depends on three groups of factors related to the seed itself :

- ➢ Internal conditions met (not dormant),
- ➢ Healthy,
- ➢ Possesses good germinative faculty and good cultural value.

3.5.1- Seeding period :

The time of sowing in the nursery is closely linked to the growing season and the planting date. In general, softwoods are sown in autumn (September, October, November, December, January at the latest) and for hardwoods (Acacia, Eucalyptus etc.) in spring (March, April, May, June at the latest). It should also be remembered that the duration of the plant's stay in the nursery must be **very short to avoid root curling.**

3.5.2- Seeding techniques :

This operation is usually carried out manually by workers from the traditional nursery. A profiled wooden plank or fingers are used to create a slight depression in the substrate to ensure that the seeds are well centered and buried at an adequate depth.

In order to optimize the production rate of pine plants in the nursery, and to ensure a success rate, the quality of the seed and the sowing date are limiting factors. Seed must be stored in a dry and cool place for the duration of the seeding period. Seeding success due to the **large quantities of seed** put in the hole and the **skilled labour** used can be thought of as a limiting factor in seeding success.

The sowing in nursery is said direct and is carried out according to several techniques:

Broadcast sowing: the seeds are distributed over the area to be sown as evenly as possible but without any precise order, -

Sowing in pots: several seeds are placed together, these small groups of seeds being regularly spaced, this technique is mainly used for sowing in place or directly in containers to obtain "clumps". --

Sowing in line: is often done in alignment, the seeds being more or less spaced, this technique is very often used for sowing in place. --

Seeding seed **by seed:** Can be considered as a variant of row seeding, each seed is deposited individually and spaced in all directions.

a. Sowing on the fly	**b.** Sowing in pots	**c.** Sowing in line

Figure 3: Seeding techniques

The sowing depth should be three times the diameter of the seed, at this depth if the conditions are favorable (adequate humidity and optimal temperature) germination would be accelerated. If the depth of the seed can exaggerate, it hinders the exit of the plant. The stay of the seedlings in the nursery depends on the size of the root ball and the location of the site.

3.5.3- Seeding advantages :

Direct sowing of seeds in the field is an attractive method for several reasons:

- ➢ It avoids nursery and transport costs;
- ➢ It eliminates any transplant trauma;
- ➢ The plants resulting from seedlings have a better shape than the plants resulting from subjects raised in nursery.

3.6- Container transport :
3.6.1- Transport of containers to the growing area :

Containers are transported mechanically with a tractor (pilot nursery), or manually by cart (traditional nursery), especially as space does not permit the movement of machinery. This step is very delicate and handling must always be done with care.

3.6.2- Installation of the containers on the cultivation area :

The containers are carefully placed on the cultivation tables. Each table is made of a galvanized steel frame capable of receiving ten of them.

3.7- Drives seedlings :
3.7.1- Watering :

This is an expensive technique. It aims to **compensate for the** soil **water deficit** during the hot season and thus to allow the young plants a **better recovery at planting** and a **better aerial and underground growth** during the first years.

3.7.1.1- Watering Quantity :

In general, the bulk of irrigation for nursery plants takes place in late spring and summer. These are also the peak (increased) periods of water use for most nursery activities as well as the periods when the Mediterranean region usually receives the least rainfall.

The frequency of watering is a function of **atmospheric conditions,** especially the **amount of precipitation recorded**. In general, watering is done every **other day in the summer** and **once or twice a week in the fall, winter and spring.**

3.7.1.2- Watering Period :

The watering period should be the period when the plant is receiving water. The application of this prescription can result in savings in the consumption of water used in the nursery:

- It is completely useless to water after a heavy rain, even when the soil is saturated with water.
- The quantity of water to be administered must be sufficient without causing asphyxiation of the plants.
- During the summer period, irrigation is done late in the evening or early in the morning.

3.7.1.3- Water Quality :

It must be slightly acidic, with a **pH less than 7, less than 550 parts/million dissolved salts**
and an electrical conductivity **(EC) less than 500 µs/cm**. In general, fairly soft and clear.

3.7.1.4- Watering techniques :

Watering can be done either by **hand** or through an **irrigation system**. Hand watering with
watering cans, hoses with spears or backpack sprayers are methods used by small nurseries. To
water containers or seedbeds: It is essential to water the seedlings with an irrigation system that
produces very fine droplets. Otherwise there is a risk that the seeds will come out or that the layer
covering them will be carried away and form a crust on the soil surface (splash effect). For this
purpose it is advisable to use a sprayer with a fine nozzle that produces almost a mist.

3.7.1.5- Watering recommendations :

In addition, this basic operation should always be entrusted to competent workers who know
when to irrigate. Stopping irrigation is recommended :

- ➢ Before the cavities are saturated with water to the point of rejection,
- ➢ Outside the hottest and most ventilated hours of the day,
- ➢ When the plants really need it,
- ➢ Depending on the temperature of the substrate in the cavities.

3.7.2- Fertilization :
3.7.2.1- Role of Nutrients :

Among the techniques likely to **provide (give) young plants a better start**, "starter" **fertilization,**
whose aim is to give, in the immediate vicinity of its roots, the elements necessary for its juvenile
growth. **Fertilizing materials** such as **fertilizers, amendments and biostimulants** ensure,
complete and **promote plant nutrition and improve the properties and life of the soil.**

Macro and micro nutrients can be distinguished, this distinction is based on the **quantities
absorbed by the** plant and **not on the importance of their role.**

3.7.2.1.1-   Macroelements or major elements :

Macroelements, are of the order of **12 elements** that represent **99% of the plant dry matter,**
required in **large quantities** by the plant to ensure its **growth and development. The
macroelements** are **Nitrogen, Phosphorus and Potassium. Secondary nutrients** are those
required in **moderate quantities** and are less likely to limit plant growth. They include
calcium, magnesium and sulphur, among others.

3.7.2.1.1.1- The effect of Phosphorus (P) :

Phosphorus is an **essential element** in the **transport of energy in the** plant, it is involved in
photosynthesis and synthesis of enzymes and proteins. It also plays an important role in **cell
division (growth) and the synthesis and transport of sugars and starches, and plant
respiration.**

3.7.2.1.1.2- The effect of Nitrogen (N) :

Nitrogen represents between **1 to 3% (DM) of** plant dry matter and up to **4 to 6%** in growing plants.

Nitrogen, as a component of chlorophyll is essential for the **growth and development of all cultivated plants,** it plays a vital role in **photosynthesis.**

Nitrogen plays a **major role in** plant **metabolism.** It is the **number 1 constituent of proteins,** essential components of **living matter, proteins and enzymes.**

3.7.2.1.1.3- The effect of Potassium (K) :

Potassium is an important constituent of plant cells. It also influences the **absorption of water by the roots and** plays a role in **plant respiration and photosynthesis.** Most crops need potassium and nitrogen in equal amounts.

Potassium, It is **mobile in the plant. It facilitates the absorption of water and proteins and allows the flowering and the fructification of the plant.** It is also an element of **plant resistance to frost, drought and disease.**

Table 1: Summary table of the importance of the macroelements assimilated by the platform

Macro-elements	Symbol	Role	Form of assimilation
Nitrogen	**N**	- Nitrogen is a fundamental element of organic nitrogen compounds such as proteins, vitamins, chlorophyll, etc. .... - Swelling action on colloids - Increases the pressure in the cells.	$NO3-$ $NO2-$
Phosphor us	**P**	- Phosphorus is an important element of various compounds such as proteins. - It participates in the activation of organic compounds - Acts negatively on the swelling of colloids.	$PO42-$
Potassiu m	**K**	- Activator of different enzymes. - Important element in the swelling of the colloids allowing the increase of the cell pressure. - Regulates water saving in the plant and reduces evaporation, thus increasing its resistance to drought. - Potassium improves the efficiency of chlorophyll assimilation and frost resistance. - Plants that are well fed with K have thicker cell walls **(increases resistance to fungal and insect attack).**	$K2+$

3.7.2.1.2-   Mesoelements or secondary elements :
3.7.2.1.2.1- The effect of Calcium (Ca) :

Calcium is an essential component of cell walls. It is also involved in the metabolism and formation of the cell nucleus. The calcium pectinate present in the cell walls provides a physical barrier to the entry of pathogens. Calcium migrates little within the plant. It is involved in plant structure, root development and fruit ripening.

3.7.2.1.2.2- The effect of Magnesium (Mg) :

Magnesium is an essential component of chlorophyll and fruit ripening. It is also involved in the synthesis of sugars, oils and fats. It promotes the absorption of phosphorus, nitrogen and sulfur.

Excessive intakes of potash can lead to magnesium deficiency. For this reason, high amounts of potash should be avoided in magnesium-poor soils.

3.7.2.1.2.3- The Sulphur effect (S) :

Sulphur is an important constituent of chlorophyll. It is involved in seed formation and nitrogen fixation. It is an essential component of several amino acids and is indispensable for protein synthesis.

Table 2: Summary table of the importance of the meso-elements assimilated by the plant

Meso-elements	Symbol	Role	Form of assimilation
Magnesium	**Mg**	- Manganese is an activator of enzymes that participate in the formation of chlorophyll, photosynthesis, protein production, etc.	Mg2+ Mg2
Sulfur	**S**	- Sulphur is a constituent element of important plant compounds, especially proteins. It also has a negative effect on the swelling of colloids.	SO42+ SO43+
Calcium	**Ca**	- Calcium is an element of some important compounds; it activates some enzymes and acts negatively on the swelling of colloids.	Ca2+

3.7.2.1.3-   The microelements :

The plant's need for trace elements is relatively low in comparison to nitrogen, phosphorus or potassium. However, their role in plant nutrition is fundamental. They act, indeed, as catalysts of several metabolic reactions. Their importance is not measured by the quantity absorbed by the crop (a few tens or hundreds of grams absorbed per hectare).

3.7.2.1.3.1- The Iron (Fe) effect :

Iron is a microelement necessary for many biochemical processes such as the synthesis of **chlorophyll**; moreover, it participates in the constitution or **activation of** a large number of **oxidation enzymes.** It also plays a role in **photosynthesis and cellular respiration.**

The leaves are, indeed, the organs of the plant which contain the major quantity of iron and in the chloroplasts one finds there the highest concentration.

3.7.2.1.3.2- The effect of Molybdenum (Mo) :

Molybdenum is a very important element that enters into the **synthesis of two enzymes** necessary for the assimilation of nitrogen: **nitrate reductase and nitrogenase.** Thus it plays an important role in the **biological fixation of nitrogen and the metabolism of phosphorus. Phosphorus favors its absorption and sulfur disadvantages it.**

3.7.2.1.3.3- The Boron Effect (B) :

Boron is used in the composition of cell walls. It has a **regulatory** action on the action of **growth hormones.** It is involved in the **growth of meristems**, the **migration of carbohydrates and assimilas, the synthesis of nucleic acids and proteins, and promotes the absorption of water.**

3.7.2.1.3.4- The effect of Zinc (Zn) :

Zinc is an important microelement for **plant growth and development.** It allows the **synthesis of proteins, starch, chlorophyll and growth hormones.** It also enters in the **composition of dehydrogenase enzymes necessary for the** synthesis of amino acids, and enters in the **synthesis of tryptophan** which is a precursor of **auxin** (hormone responsible for cell elongation). Zinc also **protects** the plant from **oxidative stress** in conditions of **strong light and drought.**

Table 3: Summary table of microelements used in fertilization

Microelements	Symbol	Role
Iron	**Fe**	- Iron is necessary for the synthesis of chlorophyll; moreover, it is used in the composition of certain enzymes.
Molybdenum	**Mo**	- Plays an important role in biological nitrogen fixation - Intervenes in the metabolism of phosphorus
Bore	**B**	- Boron is used in the composition of cell walls and carbohydrate esters. It regulates the action of growth hormones.
Zinc	**Zn**	- Zinc is an enzyme activator; it promotes the synthesis of chlorophyll and growth hormones.
Copper	**Cu**	- Copper enters the composition of different enzymes responsible for certain metabolic processes in the plant. - Copper promotes the synthesis of carbohydrates and proteins. - It also avoids an early degradation of chlorophyll: the plants keep a green and juvenile aspect longer.

Table 4: Effect of nutrients on different physiological processes

	N	P	P	Mg	Ca	B	Zn	Fe	Mo	Cu	Mn
General metabolism	▩	▩	▩	▩	▩	▩	▩	▩	▩	▩	▩
Nitrogen nutrition								▩	■		■
Growth	■		▩			■	■				
Breathing					▩			■			
Photosynthesis	▩			■			▩	■	▩	▩	▩
Sugar synthesis		▩	■	■							
Transport and accumulation of sugars					▩	■					
Other summaries (DNA lipids)					▩				■		
Resistance against illnesses			▩	▩	■					▩	
Flowering and fruit set	▩	▩	▩	▩	▩	▩	▩	▩	▩	▩	▩
Floral Induction		▩									
Flowering		■				■					
Fertilization	▩	■				■					
Nouaison						■				▩	
Production	▩	▩	▩	▩	▩	■	▩	▩	▩	▩	▩
Magnification		▩	■	▩							
Tuberization			■								
Precociousness-Maturity	▩	▩			▩						
Quality	▩	▩	▩	▩	▩	▩	▩	▩	▩	▩	▩
Product firmness		▩		■	■		▩				
Taste quality			■			■					
Conservation				▩	■	▩					

Important Elements	▩
Indispensable elements	■

It is important to know that all the mineral elements presented in the table above are naturally present in the soil and that a good maintenance of the life of the soil allows the availability of these elements for the plant.

3.7.2.4- Effect of nutrients: deficiency/excess :

3.7.2.4.1-   Macroelements :

Nitrogen deficiencies usually appear first on old leaves. Nitrogen deficiency appeared on plant leaves by a pale green or yellow coloration. Cold weather in early spring often causes temporary nitrogen deficiency in plants. This is more often the result of poor growing conditions and poor root development than a lack of nitrogen in the soil.

Phosphorus deficiency symptoms usually first appear on the tissues of old leaves. These take on a purplish red color which is often more pronounced on the underside. Severe deficiencies can also lead to dieback of the leaf tips.

Usually, the **potassium** deficiency manifests itself first on the old leaves. It can cause yellowing or marginal burning of the leaves. A severe potassium deficiency can lead to poor fruit conformation, falling flowers and young fruit.

3.7.2.4.2-   Meso-elements :

The visual symptoms of **Sulphur** deficiency are very similar to those of Nitrogen deficiency. Deficient plants are stunted (dwarfed) and have pale leaves. Sulphur deficiency can delay maturation. While nitrogen deficiency occurs first on the older leaves, sulphur deficiency occurs throughout the plant.

Magnesium deficiency first appears on the oldest leaves. A yellowing of the leaf blade between the veins is observed, which remain green. The leaves of very deficient plants have curled edges (Helix). A deficiency is considered to exist when test results indicate that the soil contains less than 20 ppm Mg.

The deficiency in **Calcium** can lead to the necrosis of the vegetative point of the plant. It can also cause the premature fall of flowers and buds.

To alleviate Calcium related tremors. The reasonable use of nitrogen helps to prevent excessive growth of vegetative organs. Good soil management practices promote root development and, as a result, water and nutrient assimilation. Well-planned irrigation also helps the plant maintain a constant supply of calcium.

3.7.2.4.3-   The microelements :

Zinc deficiency induces a reduction in the synthesis of carbohydrates and proteins.

Boron deficiency is characterized by slower growth; the youngest organs and especially the terminal buds are damaged (rot), flowers and fruits are damaged and young leaves chlorosis and dieback. Corky cracks appear on stems and roots. Its excess can lead to toxicity.

Molybdenum deficiency, induced by cold temperatures and acidic soil moisture and pH conditions. Their deficiency causes a change in leaf coloration (chlorosis of young leaves), and reduced growth and photosynthesis. In the event of very severe deficiency, the plant remains dwarfed with a progressive drying of the edges of the leaves.

Iron deficiency is manifested by interveinal leaf discoloration that begins in young leaves. It appears, especially in soils rich in active limestone or acidic soils with an excess of heavy metals.

Table 5: Nutrient Effect Table (deficiency and excess)

Macroelements				
Nutritional elements	Symbol	Excess	Carence	Form of assimilation
Nitrogen	N	- Proliferation of the vegetation part to the detriment of the flowering. - Delayed ripening; the reserve organs are of poorer quality. - The leaves are dark green. - The woody parts are less numerous. - Short, thick and less numerous roots. The roots fix the plant less well in the soil. - Not very important fruiting. - Plants more susceptible to diseases and pests.	- The plant is small, the leaves initially yellowish green to yellow become more or less orange and fall.	NO3- NO2-
Phosphorus	P	- The ability of phosphorus to be complexed in hyposoluble and insoluble forms allows substrates to have on the one hand a good phosphate reserve and on the other hand, this avoids excess phosphorus.	- Older leaves are first dark green, then reddish-purple. The stem may also take on a reddish color. - The plants are small and have a rigid appearance. At a later stage the old leaves die.	PO42-

| Potassium | K | - Plants are very fond of K and make it a luxury consumer.
- Apart from increasing susceptibility to parasites and diseases, especially virus diseases, excess K blocks the assimilation of Mg and Ca, causing signs of magnesium and calcium deficiencies even when these elements are present in sufficient quantities. | - Potassium deficiency usually appears first on the old leaves.
- The leaves are brownish green at first, then may take on a brownish red coloration.
- Chlorosis appears and develops from the edge of the old leaves, which quickly die off.
- The plants lack turgidity and wither (flaccid port). The leaves curl or roll up. | K2+ |

Mesoelements				
Nutritional elements	**Symbol**	**Excess**	**Carence**	**Form of assimilation**
Sulfur	**S**	- Excessive accumulation of salts - Disruption of molybdenum assimilation.	- Chlorosis of leaf veins and leaves ; - The leaves are less wide and the shoot has a woody appearance. - Delaying maturation - Sulphur deficiency occurs throughout the plant.	$SO_4 2+$ $SO_4 3+$
Calcium	**Ca**	- An iron antagonist, excess calcium considerably reduces its assimilation, thus inducing iron chlorosis. - Phosphorus gradually crystallizes into tricalcium phosphate ($Ca_3(PO_4)_2$).	- Calcium deficiency can lead to necrosis of the vegetative point of the plant. - Provoke the premature fall of flowers and buds.	$Ca2+$
Magnesium	**Mg**	- Marginal burn on the old leaves which become dark green.	- Magnesium deficiency first appears on the oldest leaves. - The veins of the leaves appear linear yellowing "in pearls". - Leaves fade to yellow between the veins from the base, then these patches turn brown and necrotic.	$Mg2+$ $Mg2$

Microelements				
Nutritional elements	Symbol	Excess	Carence	Form of assimilation
Copper	**Cu**	- Decreases the assimilable Fe and increases the bacterial oxidation reaction of Mn causing a deficiency of Fe and Mn. - Excess copper decreases photosynthesis; symptoms resemble those of an iron deficiency.	- Decrease in carbohydrate and protein synthesis. - Chlorosis (discoloration) and whitening of the leaf tips. - Twisting of young leaves.	Cu2+.
Molybdenum	**Mo**	- Excesses occur in organic, acidic and poorly drained soils. A slight excess can be remedied by administering calcium sulphate, which allows plants to tolerate slightly higher doses of Mo.	- Decreased growth and photosynthesis - Chlorosis of young leaves	Mo2+
Zinc	**Zn**	- This is manifested by the appearance of chlorosis accompanied by red pigmentation in the petioles and leaf veins.	- Chlorosis of young leaves - Reduction of carbohydrate and protein synthesis.	Zn2+
Copper	**Cu**	- Decreases the assimilable Fe and increases the bacterial oxidation reaction of Mn causing a deficiency of Fe and Mn. - Excess copper decreases photosynthesis; symptoms resemble those of an iron deficiency.	- Decrease in carbohydrate and protein synthesis. - Chlorosis (discoloration) and whitening of the leaf tips. - Twisting of young leaves.	Cu2+.

| Bore | B | - They can be caused by potash fertilizers containing borax as well as by irrigation with boritic water.
- As mentioned above, there is an increase in the phenomenon of guttation (**as a reminder, guttation is the exudation of water in liquid form along the leaf margin when perspiration alone is no longer sufficient).** The guttation water here contains excess boron which causes burning, browning and drying of the leaf margin. | - Slower growth
- Chlorosis and dieback of young leaves
- Terminal bud rot | Bo^{2+} |
| Iron | Fe | | - Chlorosis (discoloration) of young leaves.
- In case of acute deficiency the leaves become almost white and die back. | Fe^{2+}
Fe^{3+} |

3.7.2.5- Remedies :

Table 6: Nutrient Deficiency Remedy Table

Nutritional elements	Remedies
Nitrogen	✓ Nitrogen supply to the soil or in foliar fertilization. ✓ Drainage of soils that are too wet (reduction of denitrification). ✓ Improvement of the soil structure. Additional nitrogen supply when burying straw (C/ N ratio less than 20). ✓ Use of green manure to prevent N leaching in winter (when fall weather conditions allow).
Phosphorus	✓ Phosphate fertilization adapted to the pH. Natural slag and phosphate for acid to neutral soils and superphosphate for neutral to alkaline soils. ✓ Bring the soil in the neutral to slightly acidic pH range.
Potassium	✓ Immediate spraying of a 2% solution of potassium sulfate. ✓ Enrichment of the soil by adding potassium fertilizers (very considerable in very clayey soils). ✓ Regular potassium intake according to exports
Magnesium	✓ Magnesium contribution by kieserite. ✓ Magnesium input by dolomite (for soils with a pH below 6.5). ✓ Regular use of fertilizers containing magnesium (Patentkali, magnesium ammonium nitrate, Thomas slag, etc.). ✓ Foliar spraying of Epsom salt or other magnesium foliar fertilizers (only in the case of acute deficiencies; usually the yield is reduced).
Sulfur	✓ Use of sulphur-containing fertilizers (ammonium sulphate, simple superphosphate, potassium sulphate).
Calcium	✓ Addition of crushed limestone amendments, dolomite or defecation scum. ✓ Regular use of fertilizers containing calcium (Thomas slag and soil improvers, natural phosphates, etc.).
Iron	✓ Immediate foliar spraying of sulphate, citrate or Fe chelate. ✓ Iron chelates are added to the soil. ✓ Action on soil pH towards neutrality or low acidity. Improvement of soil organic matter content.
Molybdenum	✓ pH correction by liming. ✓ Molybdenum intake (only in case of deficiency).
Bore	✓ Correction of the pH towards neutrality. ✓ Boron supply (borax and boron-containing fertilizers). ✓ Immediate foliar spraying of boric acid.
Zinc	✓ Foliar applications of zinc salts and chelates. ✓ Lowering the pH by using acidifying fertilizer.
Copper	✓ Usually provided by fungicides containing copper. ✓ Lowering the pH by using acidifying fertilizers. ✓ Soil or foliar application of copper sulphate.

3.8-   Quality control :

Plant survival and growth in reforestation sites are strongly influenced by the quality of the plants produced in the nursery, as well as by the quality of the site (tillage, fertility and water reserves) and environmental variables (rainfall, temperature, relative humidity and vapour pressure deficit) after planting.

A plant can be considered of quality when it meets the objectives assigned to the reforestation program (protection, forest or agro-forestry production and recreation).

3.8.1- Morphological criterion :
3.8.1.1- Height-Diameter Relation :

Height and diameter are considered among the variables that can best predict plant performance after planting. Neck diameter is generally correlated with several morphological variables (height, total dry weight, root dry weight and aerial dry weight) as it is a variable that integrates the morphological response to environmental factors.

Plants with a large diameter generally have well-developed lateral roots while giving the plants a better survival rate.

Height is a good indicator of the photosynthetic capacity and the transpiration surface, which are closely correlated with the leaf area. However, height did not show a systematic correlation with survival but a good relationship with height growth.

The height category at the time of order must be met. An out-of-grade plant may be "too tall" or "too short". If more than 20% of the plants are out of the requested height category, the lot must be refused. The ratio between its height (in mm) from crown to terminal bud and its diameter at the crown (in mm) is ideal at a level of 40 to 60.

3.8.1.2- Mass ratio (aerial part / roots) and root quality :

The ratio (PA/R) is considered an index of balance between the transpiration surface (foliage) and the absorption surface of a plant (roots).

A low ratio means that roots are abundant in relation to leaf biomass and that the genus can tolerate and survive in drought conditions after planting. On the other hand, a high ratio means that roots are not abundant and that this type of plant will be more susceptible to water stress especially in semi-arid areas or in sites with high evaporative demand.

In the production of container plants, a ratio **(T/R) of 1.5 to 2 g/g** is considered satisfactory. Other work has found that the use of this ratio (T/R) remains limited in the evaluation of the quality of container-grown seedlings in predicting growth and survival in reforestation sites.

3.8.2- Physiological criteria :
3.8.2.1- Mycorhization of plants :

Mycorrhizae, are fungi proven effective in correcting iron deficiency and improving survival and growth after planting. The inoculation of the plants by **Rhizopogon** and **Pisolithus** gives the plants resistance to drought.

3.8.2.2- Nutritional status :

During the plant establishment phase, i.e. after planting, the initiation of new roots, budburst and leaf growth are influenced by the initial reserves of mineral elements. Indeed, plants with good nutritional status generally have satisfactory photosynthesis rates, an increase in leaf area and good water use efficiency.

3.8.2.3- Root growth capacity :

CCR is among the most used tests and probably the most important in the evaluation of plant quality, especially when calibrating growth substrates and crop management (irrigation and fertilization) applied in nurseries.

Root growth capacity (RGC) is the ability of a plant to initiate new roots under favourable conditions for a specified period of time. Plants with a high RAC means that their establishment in a reforestation site will be much faster.

3.8.2.4- Other quality criteria :

3.8.2.4.1-   Drying of the plant :
- **Bark:** presence of possible wrinkles and after scraping the nail, observation of poor adhesion to the stem and a state not very serous.
- **Roots:** presence on their section of brown, purplish or brown spots, signs of a dying, dead or frozen plant.
- **Softwood foliage:** foliage of the last year of incomplete and yellowing vegetation (1/3 of the needles are missing or damaged).

Figure 4: Drying of the plant

3.8.2.4.2- Health status :

Presence of damage induced by parasitic attacks (insects or fungi)

- **Insects:** defoliation or woodborers.
- **Fungus or bacteria:** anthracnose, powdery mildew, cankers, rust...
- **Rodents or game:** browsing, debarking, gnawing...

Figure 5: Parasitic attack in softwoods

3.8.2.4.3- Bud and terminal shoot :

The plant will be nonconforming if the terminal bud is absent and the shoot is not healthy. For species that form terminal buds, they must be visible and viable.

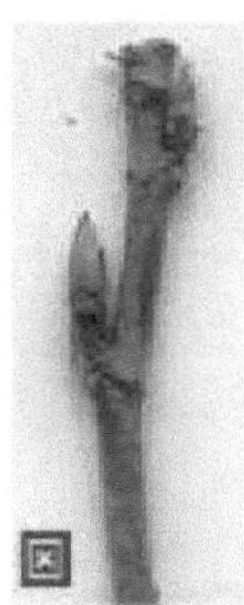

c. BT absent

b. PT faded

a. PT broken

Figure 6: Terminal bud

3.8.2.4.4- Root Defects :

The nursery culture must seek to produce plants with an abundant root system, well conformed and concentrated at the collar. Any anomaly at the root level, the plant will be discarded. Several root conformation defects are not accepted

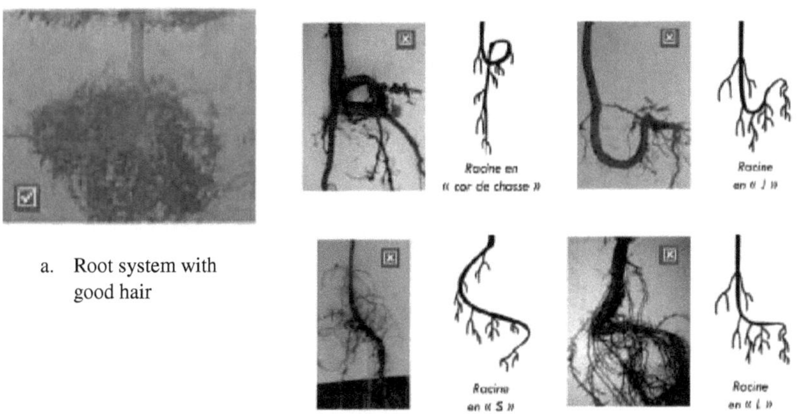

a. Root system with
good hair

b. Poorly shaped roots

Figure 7: Root defects

- Refuse any plant whose main root is severely coiled or twisted: redhibitory defects in case of coiled, twisted or deformed main roots (root in the shape of "S", root in the shape of "J", or shape of "L", whose angle is equal to or less than 100° with the stem.
- Refuse any dish that has a deficient root structure: missing rootlet or severely amputated (cut).

3.8.2.4.5- Injuries :

The plant is considered nonconforming if the removal of unhealed bark covers more than one third of the circumference of the main stem or collar or the main root. Plants with unhealed wounds will automatically be removed.

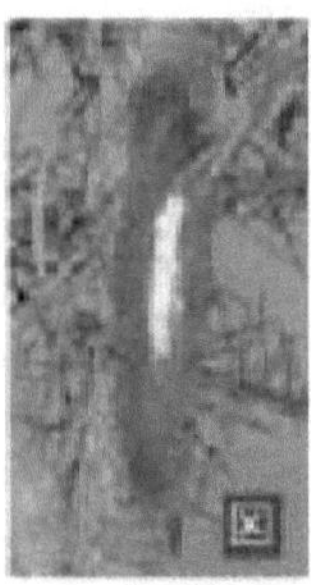
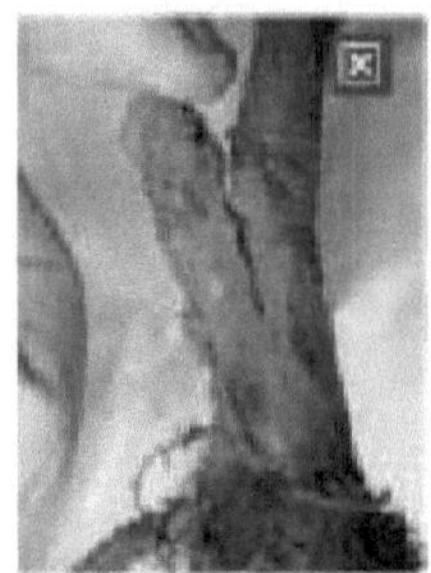

Figure 8: Injured plant

3.8.2.4.6- Plants in bucket :

For the reception of seedlings in root ball or bucket, all the criteria mentioned above obviously remain valid. However, there are also **some typical criteria for this** particular **type of culture.**

For the **reception of plants in buckets**, the agent has the possibility to analyze the rootball of one or the other "suspect" plant by **removing the substrate** for vérifier the development of the root system.

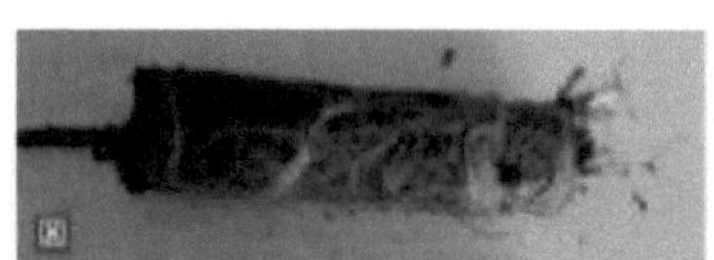

c. Root bun   **b.** Removable roots   **a.** Accepted roots

Figure 9: Bucket plants

3.8.2.4.7- Shape of stem :
- **Multiple stem:** Involves two (or more) stems starting in the first 10 centimeters above the collar.
- **Multiple Arrow:** Competing terminal branches.

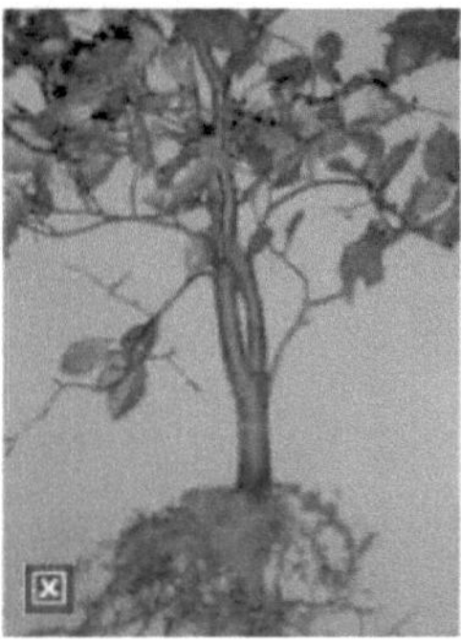

    b. Multiple stem   **a.** Multiple Arrow

Figure 10: Stem shape

3.7.2.4.1- Curvature :
- **Strong deflection:** Strong deflection of the rod occurs when the main axis deviates too much from the reference axis.
- **Curved plant:** The curvature to be taken into account can be at any level of the plant (stem, root or both). If the angle is less than 150°, the plant becomes a "plant too curved" in the defects relative to the whole. If this angle is less than 135°, the plant has in addition an individual defect.

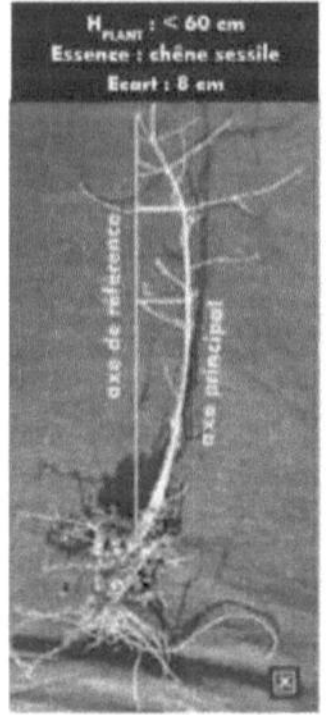

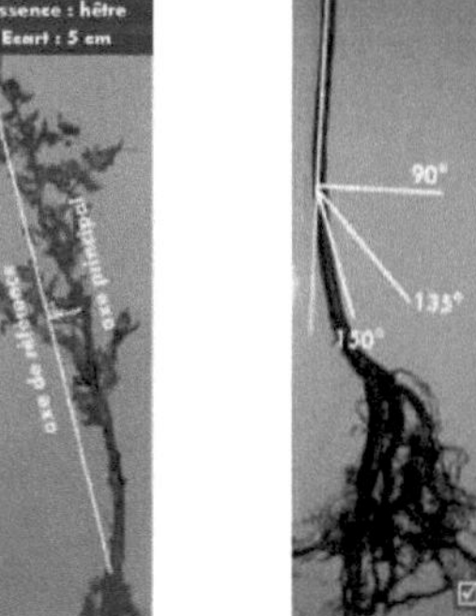

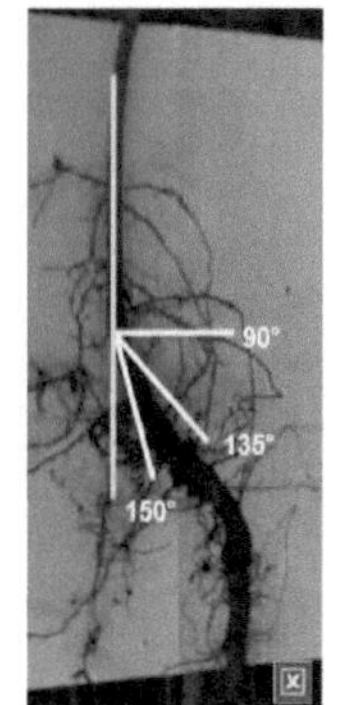

    a.   Strong deviation   b.   Curved plant

Figure 11: Curvature of the plants

3.8.2.4.8- Plant size :

The dimensions taken into consideration, are height and circumference at the collar, their combination with age makes it possible to define categories for each species. In general, for subjects of the same age, the more vigorous are those who are transplanted.

3.8.2.4.9- Controls the aerial part:

Refuse all plants with a forked top: must correct by a training pruning. If a multiple stem or a multiple flèche has been removed by a pruning, the plant is compliant for this criterion but vérifier those relating to the size of the plants

3.8.2.4.10- Tolerance margin :

Table 7: Standards for receiving forest seedlings depend on the size of the lots at vérifier.

Size of plant lots	Sample size (number of plants to be analyzed)	Number of nonconforming seedlings tolerated (after 01/07/2014)
From 1 to 50	All	3
From 51 to 100	All	5
From 101 to 2.000	100	10
From 2.001 to 10.000	200	20

Source: CDAF, 2014

3.8.2.4.11- Reception desk :

For the dishes which have a height lower than 30 cm, it is advised to measure meadows to classify the plants in height categories of 2 in 2. For those which have a height higher than 30 cm, this one is measured to 2,5 cm meadows to classify the plants in categories of height from 2,5 in 2,5 cm.

The measurement is made by placing the plant against a rigid meter, so that the lower edge of the meter is at the level of the collar. Measurements are made using a standard gauge with several notches per millimetre **(Abaque).**

The measurement is made by inserting the collar into the different notches until the compatible notch for the diameter is met.

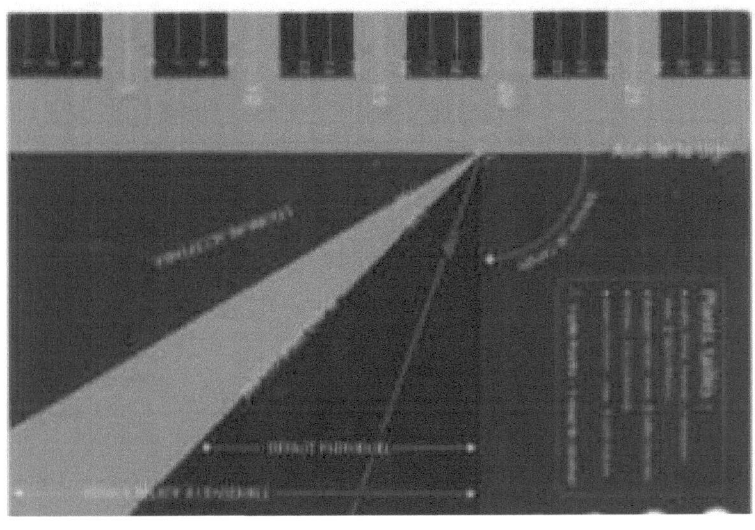

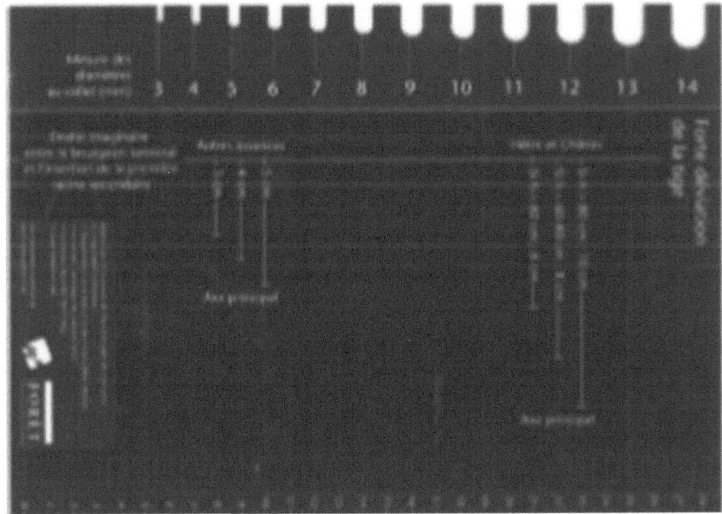

Figure 12: Receiving tray

4.   Hygiene of the plants :

Healthy plants are the goal of every nursery manager. This is not reserved for research nurseries but applies to all categories of nurseries. The hygiene of the nursery does not require the use of expensive and toxic chemicals, it can lead to a good nursery with ecological management. There are five main causes for the introduction of pathogens:

- Propagation equipment: containers, culture trays, knives, pruning shears, work surface, boxes, etc. ;
- Propagation substrates ;
- Sprinkling water ;
- Planting stock: seeds, cuttings, scions and root stocks ;
- Shoes and clothing for nursery staff and visitors.

Traditionally, there have been two basic approaches to nursery hygiene :

- ➢ **Preventive actions:** which include balanced fertilizers, use of resistant species or cultivars, timely hardening of plants, cleanliness of the nursery and training of staff ;
- ➢ **Curative actions:** which include the use of pesticides, heat, biological control or physical measures (e.g. removal of diseased parts).

5.   Interviews :

While the young plants in the nursery do not have the capacity to withstand external conditions, especially parasitic attacks, so the climatic conditions in the Mediterranean region, especially the summer period when the temperature is at its highest. The reforestation carried out there needs to be maintained to help the young plants get through the difficult period of the first years.

It is recommended to establish localized interviews around each individual in order to produce good quality seedlings. Among the interview techniques practiced in the nursery are for example :

- **Weed control:** Removal of brush and especially herbaceous **weeds in the** immediate vicinity which is considered competitive with water and nutrients.
- **Root covering:** Prevent the development of a long taproot, but also encourage the growth of a lateral root system (dryland species).
- **Seedling blight: A** disease that attacks seeds and plants and is caused by various fungi. The melting of seedlings is favoured by high humidity, wet soil surface, heavy soil, cloudy weather, excessive shade, too much plant density and alkaline conditions.
- **Marriage:** Consists of the elimination of excess seedlings to leave only the two most vigorous seedlings in the seed pot. This technique makes it possible to retain the uprooted seedlings and then transplant those with an undamaged root system.
- Transplanting: Transfer of seedlings from seedbeds to pots. This should be done as soon as possible after germination, before the roots grow as they can be damaged during the process. Transplanting is usually done with species that have very small seeds and seedlings that need special attention. When it is not done carefully, transplanting can lead to root deformities.

This maintenance can possibly be carried out from sowing to planting in the forest, until the young plant becomes resistant. After the passage from the nursery to the forest, it is very important to install a **defense** system against grazing.

The main current difficulties that prevent the success of direct seeding reside largely in the damage caused by predators (wild boar, birds, rodents ...), in addition to the main climatic hazards (frost, drought) especially for small seeds with few reserves. For this purpose it is recommended to carry out several successive replenishments to overcome the failures that have occurred, and to have an excellent success rate.

References bibliographies :

A., Seeber G. et Agpaoa. 1976. Forest tree seeds. In mannual of forestation and erosion control for the Phillipines. German Agency for Technical Cooperation-Eshborn. 1976. pp. 473-535.

B., Lérot. 2006. Les éléments minéraux. 2006.

Bhumibhamon S., 1973:. Bergen. vol.II. P.2. 1973. Sedds problems in Thailand. In "seeds Processing". Proc. Symposium Iufro Working Group on sedds problems. 1973. p. 2. Vol. II.

C.W., Goor A.Y. et Barney. 1976. Forest tree in aride zones (2 nd edition). s.l. : Ronald Press. NewYork, 1976.

CEDAF. 2014. Choice and reception of plants. Agro-Forestry Development Center of Chimay. s.l. Agroforestry in Wallonia, 2014.

D., Soltner and amélioration, 2003: . Vol.1:. 2003. The basics of plant production. s.l. Collection Sciences et techniques agricoles, 2003. Vol.1: Soil and sound.

D., Soltner. 2001. The basics of plant production. s.l. Collection Sciences et techniques agricoles, 2001. Vol.3: The plant and its improvement.

DGF. 2016. Programme d'invrestissement forestier en Tunisie. Directorate General of Forestry. Tunisia: s.n., 2016.

E.P., Bachelard. 1967. Effects of Gibberelic acid. Kietin and light on the dormant seeds of some species of Eucalyptus. s.l. : Aust.J.Bot. , 1967. p. 15.

FAO. 2004. The Forest Resources Assessment Programme. Rome-Italy: s.n., 2004.

G., Benoit D. and Alexandrian. 1979. Reforestation techniques in the Mediterranean region. s.l. Mediterranean Forests, 1979. pp. 37-42.

H., Jennick. 2006. Good cultivation practices in forest nurseries. Practical guidelines for research nurseries. 2006.

H.M., Elamin. 1975. Germination and developpemnt sudan Acacias. 1975. pp. 23-33. Sudan Sylva III Vol. 20.

J., Baeyens. 1967. Nutrition of crop plants. s.l. : Editions Nauwelaerts, 1967.

Kemp M.S., Rius A. et Wain R.L. 1975. Studies on plant growth regulating substances . 1975.

Lemhamedi M.S., Bertrand F., Luc G. and Christine G. 2006. Guide pratique de production en hors sol de plants forestiers, pastoraux et ornementaux en Tunisie. 2006.

M., Evenari. 1957. The physiological action and biological importance of germination inhibitors. Univertity of Cambridge. 1957. pp. 21-43.

M.S., Dourossin. 1987. Research on some containers used in nurseries in Burkina Faso. University of Rural Development. 1987.

MAAAF. 2014. Guide to Nursery and Ornamental Plants. Cultivation and Integrated Pest Management. Ministry of Agriculture and Food and Business. Toronto-Canada: n.d., 2014.

—. **2009.** Macroelements and secondary nutrients. LI culture. Ministry of Agriculture, Food and Rural Affairs. Ontario-Canada: n.n., 2009.

ITTO. 2002. ITTO Guidelines for the Restoration, Management and Rehabilitation of Degraded and Secondary Tropical Forests. Yokohama: s.n., 2002. Forest Policy Vol. 13. ITTO.

P., Morard. 1995. Above-ground vegetable crops. s.l. : Publications Agricoles, 1995.

Presowing treatment of seeds to speed germination. In seeds of woody plants in the USA. **Bonner F.T., Mc Lemore B.F. et Barnett J.P. 1974.** Wachington D.C : s.n., 1974, Agriculture Handbook N°450.For. Services, USDA.

R.L., Wiliam. 1992. A guide to handling forest seeds in the special case of tropical regions. Danida Forest Seed Center. Food and Agriculture Organization of the United Nations, FAO Forestry Paper 20/2. Rome-Italy: s.n., 1992.

S., Dominique. 2003. he basics of plant production. s.l. Collection Sciences et techniques agricoles, 2003. Vol. 1: Soil and soil improvement.

S., Markku. 2009. Towards a definition of forest degradation: A comparative analysis of existing definitions. Forest Resources Assessment Programme. 2009.

V.L., Balleux P. and Phillipe. 2004. Quality, conditioning and control of forest plants. Cahier technique n°28. Institut pour le Developpemenr Foresie. 2004.

Web references :

- http://www.agrisymbiose.com/app/agrisymbiose/files-module/local/documents/green_book.pdf
- http://www.virtual.chapingo.mx/dona/paginaIntAgronomia/deficiencias.pdf
- https://fertilisation-edu.fr/nutrition-des-plantes/le-role-des-elements-nutritifs/oligo-elements.html
- https://www.bio-enligne.com/fertilisation/199-oligo-elements.html
- http://219.127.136.74/live/Live_Server/154/ps13e.pdf
- http://www.fao.org/3/T0122F/t0122f06.htm
- http://www.fao.org/3/T0122F/t0122f00.htm
- http://www.fao.org/3/T0122F/t0122f07.htm
- http://www.fao.org/3/ad232f/ad232f11.htm

APPENDIX

Primary symptom on the leaves	Ages of Leaves Affected	Other possible symptoms	Deficiency/excess
Chlorosis	**Elderly then young people**	- The leaves take on an orange color. - Pronounced veins. Straight and stiff leaves	**Deficiency N**
	Youth	- Winding and/or embossing of sheets. - Brown coloration of the leaves	**Deficiency Ca :** **(Excess K)**
		- Shrunken plant Similarity N deficiency	**Deficiency S**
		- Chlorosis of the blade (yellow to white) followed by veins. - Necroses	**Deficiency Fe :** **(Excess Ca) ;** **(Excess Cu)**
		- Short internodes rosette of leaves.	**Deficiency Zn :** **(Excess Ca)**
		- Red pigmentation on petiole and veins.	**Excess Zn**
	Elderly	- Necroses.	**Acute deficiency P**
		- Internervial chlorosis, Brown necrosis.	**Mg deficiency:** **(Excess K);** **(Excess Ca)**
		- Chlorosis by range. Browning and necrosis on the edges and then the veins.	**Deficiency Mn:** **(Excess Ca);** **(Excess Cu)** **Excess Mn**
		- Wilting of leaf edges. Necrosis. Malformations/deformations of the leaf blade, stems, petioles.	**Deficiency Mo :** **Excess S**

Bronze coloring sheets	**Elderly**	- Wilt - Necroses - Leaf edge rippling	**Deficiency K**

Stain discoloration.	**Youth**	- Deformation/increases. - Browning/blackening. - Cracking on petioles and stems	**Deficiency B:** (Excess Ca)

Burns	**Elderly**	- Marginal burns	**Excess Mg**
	Youth	- Browning, - Drying of the edge. - Guttation - Burns,	**Excess B**

Purple coloring	**Elderly**	- Bluish green leaves + yellowing - Rippling of the end of the leaves. - Drying out.	**Benign deficiency P**
	Youth	- Red coloring on veins and petioles. Chlorosis	**Excess Zn**
Orange color	**Elderly then young**	- Chlorosis - Pronounced veins. - Straight and stiff sheets	**Deficiency N**

Dark green tending to bluish	**Young and old**	- Beach chlorosis on young leaves. - Rolling, embossing, malformation of the foliage.	**Cu deficiency: (Excess Ca)**
	Elderly	- Yellowing - Purple coloring, undulation of the end of the leaves followed by a drying.	**Benign deficiency P**

Strong guttation	**Youth**	- Burns, - Browning, - Drying of the edge.	**Excess B**

Wilt	**Elderly and very quickly the young.**	- Growth retardation. - Bronze coloring of the foliage. - Necroses. - Rippling of the leaf edge.	**Deficiency K**

Dark green leaves	**Elderly**	- Leaves tending to bluish with yellowing	**Beginner's deficiency P**
		- Purple coloring, undulation of the end of the leaves followed by a drying.	
		- Marginal burns	**Excess Mg**
	Young and old	- Short and thick roots.	**Excess N**
		- Flowering and fruiting almost non-existent.	

Printed by Books on Demand GmbH, Norderstedt / Germany